Cell Free Systems Conference 2019

Boston, Massachusetts, USA
4 – 6 December 2019

ISBN: 978-1-7138-0571-7

Printed from e-media with permission by:

Curran Associates, Inc.
57 Morehouse Lane
Red Hook, NY 12571

CURRAN ASSOCIATES INC.
proceedings.com

Some format issues inherent in the e-media version may also appear in this print version.

Printed with permission by Curran Associates, Inc. (2020)

For permission requests, please contact AIChE
at the address below.

AIChE
120 Wall Street, FL 23
New York, NY 10005-4020

Phone: (800) 242-4363
Fax: (203) 775-5177

www.aiche.org

Additional copies of this publication are available from:

Curran Associates, Inc.
57 Morehouse Lane
Red Hook, NY 12571 USA
Phone: 845-758-0400
Fax: 845-758-2633
Email: curran@proceedings.com
Web: www.proceedings.com

TABLE OF CONTENTS

PROTOTYPING AND CHARACTERIZATION

KEYNOTE 3

MANUFACTURING

POSTER SESSION

Freeze-Dried, Cell-Free Synthetic Biology

Cell Free Systems Conference

Authors:

Collins, J. J., Massachusetts Institute of Technology

Synthetic biology is bringing together engineers, physicists and biologists to model, design and construct biological circuits out of proteins, genes and other bits of DNA, and to use these circuits to rewire and reprogram organisms. Recent efforts have shown that it is possible to freeze-dry synthetic gene circuits and cell-free transcription-translation systems onto paper and other substrates.

Here we describe how these technologies are being used to create rapid diagnostics, on-demand biomolecular manufacturig capabilities, synthetic biology educational kits, and next-generation wearables, enabling the power and diversity of biology to be harnessed outside of the laboratory.

Using Cell-Free Systems to Develop and Deploy Biosensors That Meet Army Needs

Cell Free Systems Conference

Authors:

Buckley, P., CCDC Chemical Biological Center

Cell-free protein expression systems are inherently useful for developing fieldable biosensors because of their ability to maintain cellular functions without the requirements of living organisms. Sensors are needed in the field to detect a multitude of different targets that allow for the recognition of target molecules and subsequent activation of reporter molecules. The Department of Defense has a strong interest in utilizing cell-free reactions to help alleviate logistical burdens associated with traditional methods of detection. Designing biosensors to have low size, weight, and power requirements often requires the adaptation of assays into a paper-based, eye-readable format that is simple enough to be used in the field. Furthermore, there has been a significant push in the government to have independent testing to help validate the applicability of designed assays. The variety of lysates and circuits for detection available require vigorous testing and characterization in order to determine their feasibility for use in deployable sensors.

Metabolic Perceptrons for Neural Computing in Cell Free Systems

---

Authors:

Kushwaha, M., Institut national de la recherche agronomique (INRA)

Pandi, A., INRA

Koch, M., INRA

Voyvodic, P., Centre de Biochimie Structurale

Soudier, P., Genoscope

Bonnet, J., Centre de Biochimie Structurale

Faulon, J. L., Genoscope

Synthetic biological circuits are promising tools for developing sophisticated systems for medical, industrial, and environmental applications. So far, circuit implementations commonly rely on gene expression regulation for information processing using digital logic. Here, we present a different approach for biological computation through metabolic circuits designed by computer-aided tools, implemented in both whole-cell and cell-free systems. We first combine metabolic transducers to build an analog adder, a device that sums up the concentrations of multiple input metabolites. Next, we build a weighted adder where the contributions of the different metabolites to the sum can be adjusted. Using a computational model fitted on experimental data, we finally implement two four-input perceptrons for desired binary classification of metabolite combinations by applying model-predicted weights to the metabolic perceptron. The perceptron-mediated neural computing introduced here lays the groundwork for more advanced metabolic circuits for rapid and scalable multiplex sensing in cell-free systems.

Cell Free Operating Systems for Diagnostics

Authors:

Ellington, A., University of Texas at Austin

Bhadra, S., University of Texas at Austin

Nguyen, V., The University of Texas at Austin

Torres, J. A., The University of Texas at Austin

Kar, S., University of Texas at Austin

Wang, S., The University of Texas at Austin

Strand exchange nucleic acid circuits are an example of a complex computational paradigm that can be readily developed in a cell free environment. While some strand exchange circuits can operate even in the absence of enzymes, replication, transcription, and translation can be readily adapted to participate in computations in a cell-free format. In consequence, we have developed strand exchange circuits that carry out multi-layered computations on amplified nucleic acids, and that can be readily used at point-of-care. Based on our work, we have developed working diagnostics that can detect small numbers of nucleic acid analytes for a wide variety of disease states with little or no upstream purification, and will report progress in distributing these in educational and commercial settings, in the United States and in partnership with institutions in Africa. One of our greatest successes is a diagnostic that can in a single reaction calculate and output the presence of virtually any Zika virus variant, while excluding related viruses. As an adjunct to these approaches, we have begun to turn the traditional paradigm for cellular biosensors on its head by using cells solely as factories for the production of reagents, and integrating lyophilized cells directly with cell-free strand exchange circuitry: in this model, cells become otherwise inert bags of enzymes that serve as inputs for nucleic acid devices, rather than the other way around. As these utility of these methods continues to expand, we have found that nucleic acid circuits can be "stacked" into functional networks, and are exploring cell-free neural net classifiers for disease states.

A Measurement Foundation for the Future of Cell-free Systems

Cell Free Systems Conference

Authors:

Romantseva, J., Material Measurements Laboratory, National Institute of Standards and Technology

Cell-free systems have emerged as powerful tools for the expression of recombinant proteins. However, a lack of reproducibility persists within and across laboratories engaged in research using these systems. Measurement tools and methods specific to cell-free systems are essential to overcome this challenge, supporting intra- and interlaboratory collaboration, assisting nascent industry, and improving fundamental understanding of biology. The National Institute of Standards and Technology (NIST) seeks to foster innovation and build confidence in quantitative biology and biomaterial measurements. As part of a broader investment in synthetic and engineering biology, NIST is building technical competence and measurement capabilities in cell-free systems to deliver measurement assurance and measurement science for these systems. Towards this end, NIST is investing in automation to advance the utility and lower the barrier to adoption of cell-free systems.

Tackling Global Health Challenges with Cell-Free Synthetic Biology: New Technologies for Rapid, on-Demand Water Quality Monitoring

---

Authors:

Lucks, J. B., Northwestern University

Recent advances in cell-free synthetic biology have enabled the development of powerful new diagnostic technologies. These technologies leverage the ability of cell-free systems to support natural and engineered biosensing mechanisms to create detectable signals when target analytes are present. The ability to freeze-dry these systems for long-term storage and distribution prior to rehydrating with test samples has enabled the development of rapid, low-cost and programmable point-of-use diagnostics for a range of important targets of global health concern, with particular successes in nucleic acid diagnostics for pathogens and human health biomarkers. Here, we present our work on expanding the reach of cell-free, synthetic biology-enabled diagnostics by developing several platforms for the detection of toxic contaminants - including heavy metals, halides, drugs, pesticides, and personal care products - in aqueous samples. Our technologies combine several new innovations in biosensor programming, including protein and RNA-gated inducible regulators that sense a range of targets, RNA-level outputs for enhanced detection speed, downstream genetic circuit computation for signal processing, and coupling to metabolic bioconversion to expand the sensing repertoire. We have also partnered with global health and water experts to field-test these platforms by detecting both heavy metal contamination caused by wildfires in Paradise, CA as well as endemic fluoride contamination of freshwater in Costa Rica. Collectively, our work is enabling the transition of technologies from the lab into the real-world where they are needed most and will help achieve progress towards the United Nations - Sustainable Development Goal "ensuring access to clean water across the globe. We hope that lessons learned will help the community to demonstrate the power of cell-free synthetic biology to solve major global challenges.

Expanding the Scope of Cell-Free Diagnostics: Equipment-Free Quantification of Biomarkers in Human Serum

Cell Free Systems Conference

Authors:

McNerney, M., Georgia Institute of Technology

Styczynski, M. P., Georgia Institute of Technology

Zhang, Y., Georgia Institute of Technology

Steppe, P., Georgia Institute of Technology

Silverman, A., Northwestern University

Jewett, M. C., Northwestern University

Cell-free expression (CFE) systems have enabled a variety of low-resource diagnostic tools, but these have been largely limited to assays that detect (rather than quantify) a target analyte. Biomarker quantification is challenging in CFE diagnostic systems, largely because the complex makeup of biological samples affects assay output, which precludes the use of traditional calibration methods.

We will present the development of a CFE diagnostic platform that can quantitatively measure biomarkers in complex matrices in an equipment-free fashion. The key innovation is the development of a standardization approach that allows equipment-free quantification in assays consisting of 25-100% human serum. The standardization scheme is designed so that the serum sample to be analyzed is added both to the test reaction and to a set of standard reference reactions. Analogous to a litmus test, the color of the test can be matched to the colors of the reference reactions so that the user can easily determine the sample concentration. We demonstrate this method's utility, reliability, and robustness with a biomedically relevant zinc sensor. The test reliably quantifies clinically relevant zinc concentrations and is a significant step towards a low-cost, field-deployable test for zinc deficiency, which causes an estimated 100,000 annual childhood deaths. We demonstrate the generalizability of our quantification method by applying it to a sensor that uses an established toehold switch. The developed test is the first CFE assay to provide equipment-free biomarker quantification in a complex matrix, and the platform developed can be readily expanded to quantify other molecules in cell-free expression reactions.

Identification of Additive Formulations for Enhanced Shelf-Life and Productivity of E. coli Cell Extracts

---

Authors:

Oza, J. P., California Polytechnic State University, San Luis Obispo

Cell-Free Protein Synthesis (CFPS) is a platform biotechnology that enables a breadth of applications. However, field applications remain limited due to the poor shelf-stability of aqueous cell extracts required for CFPS. Lyophilization of *E. coli* extracts improves shelf-life but remains insufficient for extended storage at room temperature. To address this limitation, we mapped the chemical space of ten low-cost additives with four distinct mechanisms of action in a combinatorial manner to identify formulations capable of stabilizing lyophilized cell extract. We report three key findings: 1) unique additive formulations that maintain full productivity of cell extracts stored at 4ËšC and 23ËšC; 2) additive formulations that enhance extract productivity by nearly 2-fold; 3) a machine learning algorithm that provides predictive capacity for the stabilizing effects of additive formulations. These findings provide a simple and low-cost advance toward making CFPS field-ready and cost-competitive for biomanufacturing.

INSPECTR - A breakthrough platform for cell-free molecular diagnostics

Cell Free Systems Conference

Authors:

Carlson, P., Sherlock Biosciences

There is an unmet need for low-cost, simple-to-use, shelf-stable nucleic acid diagnostics. Despite the growing market size and global impact of molecular diagnostics, there remains a gap in providing this technology directly to individuals in a format that is as cost effective and as simple to use as a lateral flow immunoassay. Existing molecular diagnostic systems highlight a key tradeoff between cost and simplicity, with higher cost instruments and consumables required to enable a simpler workflow with improved usability. There are currently no FDA-approved over-the-counter molecular diagnostic tests, as current nucleic acid detection technologies struggle to achieve the price, usability, and deployability criteria for such a test.

Sherlock Biosciences has developed a platform for cell-free molecular diagnostics called INSPECTR to address this need. INSPECTR (Internal Splint-Pairing Expression Cassette Translation Reaction) leverages cell-free systems for sensitive detection of nucleic acids at ambient temperature with single nucleotide resolution. The highly programmable nature of INSPECTR enables the flexibility to link the detection of any nucleic acid input with any nucleic acid or protein output. We have demonstrated that INSPECTR can be coupled to an inexpensive lateral flow readout, serving as the first demonstration instrument-free, amplification-free, ambient temperature nucleic acid detection. Our breakthrough platform combines the specificity, rapid development, and broad applicability of a molecular diagnostic with the low-cost, stability, and direct-to-user applicability of lateral flow immunoassays.

Persia for Direct Fluorescence Measurements of Transcription,Translation, and Enzyme Activity in Cell-Free Systems

Cell Free Systems Conference

Wednesday, December 4, 2019 - 1:50pm-2:15pm

Authors:

Carr, P., MIT Lincoln Laboratory

Walsh, D. III, MIT-Lincoln Laboratory

Wick, S., MIT Lincoln Laboratory

Bobrow, J., MIT Lincoln Laboratory

Thorsen, T., MIT Lincoln Laboratory

Kong, D., MIT

Hamad-Schifferli, K., University of Massachusets

Mroszczyk, K., MIT Lincoln Laboratory

In a cell-free reaction, we would like to know what is happening and when it is happening. We employ an approach we term PERSIA (for PureExpress-ReAsH-Spinach-In-vitro-Analysis) that allows us to observe the core processes of cell-free transcription and translation during the reaction. We combine small molecule labels (DFHBI and ReAsH) with RNA and peptide tags (Spinach and tetracysteine) that produce fluorescent signals upon bindingreporting on the synthesis of the full-length biomolecule. Avoiding GFP and other fluorescent proteins allows us to measure production directly for our many different proteins of interest. Coupled with de novo DNA synthesis, we have applied PERSIA to answering several different kinds of questions, such as the practicability of designed genetic codes, scanning mutagenesis of full length proteins, and probing drug resistance mutations. Additional opportunities with PERSIA remain such as optimizing protein yields for heterologous expression.

Accessible and Easy-to-Use Educational Tools to Teach Molecular and Synthetic Biology Using Freeze-Dried, Cell-Free Technology

Cell Free Systems Conference

Authors:

Huang, A., miniPCR

Hands-on demonstrations greatly enhance the teaching of STEM concepts and foster engagement and exploration in the sciences. While numerous chemistry and physics classroom demonstrations exist, few biology demonstrations are practical and accessible due to the challenges and concerns of growing living cells in classrooms. Here I introduce a platform to develop hands-on molecular and synthetic biology educational activities based on easy-to-use, shelf-stable, freeze-dried, cell-free (FD-CF) reactions, which are simply activated by water. By using fluorescent proteins as a visual output, I created a variety of engaging modules using this platform that can teach the central dogma of biology, how certain cellular functions work, and other basic molecular biology topics that are otherwise difficult to easily teach in a hands-on manner. By expanding the platform to other non-visual outputs (such as smell or touch), as well as further incorporating components, such as RNA switches, I also developed modules that can teach more advanced biology topics, such as biochemistry, biomaterials, and synthetic biology, as well as basic laboratory skills such as pipetting, experimental design, and the scientific method. Pilot testing of a prototype kit based on these elements were tested in classrooms across the country and initial results suggest that the activities are accessible, easy to use, educational, and engaging for high school students. Overall, the platform introduces low-cost, user-friendly, and hands-on activities that can be used in classrooms to improve the quality of biology education and open the door for student-driven, independent explorations in the life sciences.

Rapidly Prototyping CRISPR Biology and Technologies with an All-E. coli Txtl System

Cell Free Systems Conference

Authors:

Beisel, C. L., Helmholtz Institute for RNA-based Infection Research

CRISPR-Cas systems represent widespread adaptive immune systems in bacteria and archaea that have become versatile technologies driving applications in gene therapy, diagnostics, antimicrobials, functional genetic screens, and more. The vast diversity of these immune systems and their broad applicability as technologies have spurred efforts into their discovery, characterization, and implementation. However, traditional cell-based and in vitro techniques represent a notable bottleneck. Here, I will describe how cell-free transcription-translation (TXTL) systems can be employed to greatly accelerate the prototyping of CRISPR biology and technologies. I will show how an all-*E. coli* TXTL system can be used to rapidly and scalably characterize target recognition by CRISPR nucleases, elucidate rules for guide RNA design, and identify different inhibitors of nuclease activity. I will also illustrate the power of TXTL using two examples from our group. In one example, we developed RNA-responsive guide RNA 'switches' that only drive nuclease activity in the presence of a selected RNA sequence. We used TXTL to devise general design rules for these switches, allowing us to apply these rules to generate functional switches that respond to synthetic and native RNAs in *E. coli*. In another example, we developed two means to rapidly reverse gene repression by catalytically-dead CRISPR nucleases through anti-CRISPR proteins or upstream-binding guide RNAs. TXTL allowed us to elucidate design rules for either means and translate insights into orthogonal repression and de-repression circuits as well as dynamic feedback controllers. Overall, these examples highlight an important and growing use of the ease, speed, and scalability of TXTL.

Artificial Cells on a Chip

Cell Free Systems Conference

Authors:

Bar-Ziv, R., Weizmann Institute of Science

Daube, S. S., Weizmann Institute of Science

Divon, Y., Weizmann Institute of Science

Falkovich, R., Massachusetts Institute of Technology

Greiss, F., Weizmann Institute of Science

Levy, M., Weizmann Institute of Science

Mostov, P., Weizmann Institute of Science

Ricouvier, J., Weizmann Institute of Science

Stern, N., Weizmann Institute of Science

Vonshak, O., Weizmann Institute of Science

We study the assembly of artificial cells on a chip from the single cell level to multicellular architecture and communication. We will describe recent progress toward programmable synthesis and assembly of cellular machines, self-replication, memory transactions, single-molecule fluctuations, synchrony and pattern formation, as well as electric field control of gene expression.

Cell-free Cells

Cell Free Systems Conference

Authors:

Adamala, K., University of Minnesota

All of biological science to date has been done on a sample size that equals one: that of the modern, terrestrial life. The development of cell-free transcription and translation systems allows us, for the first time, to expand the chassis for biological experimentation. Cell-free protein expression can be used to create analogues of various biological processes, reconstituting natural functionalities and building new ones. We can include variety of unnatural biochemistries, and explore physical and chemical limitations of the life as we know it.

Learning how to engineer a novel live cell from set of completely understandable parts will provide a leap in our understanding of life, and will also result in novel technologies and tools beneficial for the scientific community and society. Tools developed for synthetic cells can be used for molecular biology and basic research, as well as for biotechnology and biomedical applications.

Cell-free systems already allow us shortening of the classical design - build - test cycle of biological experimentation. Encapsulating cell-free translation machinery inside a liposome - creating a synthetic minimal cell, allows adding a layer of complexity that turns cell-free systems into experiments that approach in complexity simplest known live cells.

We develop new tools for protein evolution and for reconstitution of natural processes in cell-free systems and in synthetic cells, investigating the limits of natural life and the possibilities of biological engineering outside of the natural chassis.

A Direct Gene Circuit - Electronic Interface for Cell-Free Synthetic Biology

Cell Free Systems Conference

Authors:

Pardee, K., University of Toronto

Sadat Mousavi, P., University of Toronto

Smith, S., Bucknell University

Chen, J., University of Toronto

Karlikow, M., University of Toronto

Tinafar, A., University of Toronto

Robinson, C., University of Toronto

Liu, W., University of Toronto

Ma, D., Arizona State University

Green, A. A., Arizona State University

Kelley, S., University of Toronto

Cell-free systems are emerging as key platforms for application-driven synthetic biology. Unlike their cell-based counterparts, cell-free systems can be assembled without the laborious task of moving engineered elements into cells or the balancing of their function within the context of complex endogenous molecular programs. While potential applications for cell-free systems are wide-ranging, their biosafe format makes them especially well-suited for de-centralizing sensing and diagnostics. To date, gene circuit-based sensors have primarily used optical proteins (e.g. fluorescent, colorimetric) as reporter outputs. This has limited the potential to measure multiple distinct signals. Here we present an electrochemical interface that allows measurement of multiple distinct reporter outputs within the same volume. To achieve this, we have developed a scalable system of reporter enzymes that convert sensor activation into sequence-specific cleavage of DNA. Upon liberation of the resulting ssDNA strands, an electrochemical signal is generated through base pairing at the surface of nanostructured microelectrodes coated in complementary DNA. We will discuss the development of this electrochemical interface and share proof-of-concept work that showcases a small molecule-regulated genetic switch and toehold switch-based sensors for the detection of resistance genes against the last line antibiotic colistin. We see this technology as an exciting new medium for the field of synthetic biology and one with the potential to drive further interdisciplinary interactions with materials, hardware and software.

Cell-Free Gene Expression in Synthetic Multicellular Structures and Organelles

---

Authors:

Simmel, F. C., TU Munich

Compartmentalization of cell-free gene expression reactions is a common strategy towards the realization of artificial cellular systems. Based on this previous work, we have recently investigated a variety of approaches to create multi-compartmentalized structures to investigate molecular communication and division of labor within such systems.

For instance, we utilized quorum sensing signals to establish communication between compartmentalized gene expression systems and bacteria, or other small molecule signals

to realize signaling within artificial, tissue-like structures made with the droplet-interface-bilayer technique. A different type of compartmentalization was investigated using genetic templates immobilized in gel particles, which were utilized to separate transcription and translation processes into separate compartments, or 'artificial organelles'.

Using such and similar systems, we hope to be able to engineer bottom-up synthetic biological systems of higher complexity, with the ability to functionally differentiate depending on environmental cues.

Integrated Mesoscale Synthetic Regulatory Networks Built from Genelets

Authors:

Schulman, R., Johns Hopkins University

Schaffter, S., Johns Hopkins University

Cells are capable of orchestrating complex behaviors, including differentiation and environmental adaptation, using genetic regulatory networks (GRNs) consisting of interconnected genes that regulate one another. Synthetic in vitro chemical reaction networks that mimic cellular GRN dynamics might direct synthetic molecules to exhibit some of the complex dynamic interactions that underlie these complex behaviors. A simple yet powerful toolbox for assembling synthetic GRN analogs are in vitro transcriptional circuits. These circuits comprise short synthetic genelets that are transcribed by viral RNA polymerases and use short nucleic acid sequences to regulate genelet expression, making them straightforward to program and implement. I will describe an updated genelet toolbox that minimizes component crosstalk to enable the construction of large multi-functional regulatory networks. The toolbox comprises >10 unique genelet regulatory motifs that can be interconnected to rapidly create different mesoscale synthetic regulatory networks. Using predictive kinetic models to guide network design and implementation, we assemble genelet networks that (1) exhibit switchable multi-stability, (2) execute temporal programs inspired by cellular developmental pathways, and (3) integrate the behaviors of 1 and 2 to orchestrate state-specific signal expression programs. I will also highlight key challenges that must be addressed to use these networks in downstream applications, including fuel consumption/management and material coupling.

Geometric Control of Bio-Machine Assembly By a Genetic Program on a Chip

Authors:

Daube, S. S., Weizmann Institute of Science

Levy, M., Weizmann Institute of Science

Falkovich, R., Massachusetts Institute of Technology

Vonshak, O., Weizmann Institute of Science

Divon, Y., Weizmann Institute of Science

Bar-Ziv, R., Weizmann Institute of Science

From viruses to ribosomes, large macromolecular machines self-assemble inside cells in a highly orchestrated and efficient manner, regulated by the spatial organization of genes in operons and clusters, by gene expression foci and by the crowded cytoplasm. We seek to define a cell-free architecture of DNA programs to achieve geometric self-organization of protein assembly pathways. We present an emerging surface approach to recreate and investigate bio-assembly processes crucial to the creation of a self-replicating artificial cell. RNA and protein machine parts are synthesized in a cell-free reaction from high-density surface-immobilized gene compartments and self-assemble into large complexes on nearby surface traps. The DNA compartments localize the gene-expression machinery, creating a local source of RNAs and proteins, coupling synthesis to assembly. Surface trapping of assembly products provides structural scaffolding and a highly sensitive fluorescent imaging in time and space. Using this methodology, we recreate the assembly pathway of both T4 bacteriophage baseplates and *E. coli* 30S ribosomal subunits starting from only their genes, realizing a cell-free scenario for ribosomes making ribosomes.

Engineering Synthetic Organelles to Spatially Organize Biochemical Reactions in Artificial Cells and Cell-Free Systems

Cell Free Systems Conference

Authors:

Niederholtmeyer, H., University of California, San Diego

Bhattacharya, A., University of California San Diego

Devaraj, N. K., University of California, San Diego

Living cells have evolved numerous mechanisms to spatially organize reactions in their interior. In fact, compartmentalization is a defining characteristic of eukaryotic cells with their membrane-bound and membraneless organelles. By colocalizing specific reactions and separating others, synthetic organelles have the potential to improve the biochemical capabilities of artificial cells and cell-free systems. Simplified and engineerable models of natural cellular organelles will also lead to a better fundamental understanding of the effects of spatial organization on biochemical reactions. We have recently reported porous polymer cell-mimics with artificial nucleus-like compartments. Artificial nuclei composed of a clay-DNA hydrogel contain the templates for transcription and translation, which program cell-mimics to communicate with their neighbors. In addition, we are currently developing an artificial organelle system that consists of liquid, phase-separated lipid sponge droplets formed by a synthetic glycolipid. Droplets can sequester both hydrophobic and hydrophilic molecules, and in contrast to membrane-bound vesicles, encapsulated cargo remains in contact with the surrounding medium. Highly-efficient sequestration of specific proteins into droplets can be programmed by affinity tags, and we use this property to study the effects of colocalization and separation on enzymatic reactions. For instance, light-responsive release and encapsulation of proteins allows us to dynamically control ClpXP-mediated protein degradation by switching between proteolysis and protection of the substrate. Our results suggest that artificial organelle structures built from synthetic materials facilitate the engineering of spatial organization and will help integrate increasingly complex functions in artificial cells and cell-free systems.

A Cell-Free Protein Synthesis Platform for Rapid and Quantitative Protein-Protein Interaction Analysis

Cell Free Systems Conference

Authors:

Hunt, A., New Mexico Institute of Mining and Technology

Jewett, M. C., Northwestern University

Protein-protein interactions (PPIs) facilitate many core biological functions, from signaling to complex and compartment assembly. Despite their importance and ubiquity, the study of PPIs is still often laborious and time-consuming. Additionally, there are few methods for the investigation of PPI kinetics in high throughput, despite the fact that understanding interaction kinetics is often critical for understanding and designing biological function. In this work, we present a robust cell-free method for the characterization of PPIs that enables high throughput, rapid, and quantitative characterization of both PPI interaction networks and kinetics. Our platform leverages cell-free protein synthesis, an Echo® 525 acoustic liquid handler, and a split luciferase complementation reporter to enable rapid expression and characterization of tens to hundreds of interaction partners on short time scales. We will present our results to validate the platform and characterize novel PPIs. Specifically, we will validate the method by measuring the kinetics of the well-studied PUMA and MCL-1 model PPI system without purification. Additionally, we will show our efforts to validate the platform's ability to characterize novel PPIs by examining tens of novel *de novo* designed heterodimers. A key feature of our approach is the ability to measure both the (1) interaction network and (2) kinetics of all strongly interacting and stable heterodimers within a single platform. Our results show that cell-free protein synthesis coupled with liquid handling robotics is a powerful tool for the rapid and high throughput characterization of PPIs. We expect this method will accelerate both the study and engineering of PPIs.

Boosting the Activity of CHO-Based Cell-Free Protein Synthesis Factories for High-Throughput in Vitro Production of Functional Antibodies

Cell Free Systems Conference

Authors:

Heide, C., Imperial College London

Buldum, G., Marmara University

Ces, O., Imperial College London

Kontoravdi, C., Imperial College London

Polizzi, K., Imperial College London

Cell-free protein synthesis (CFPS) is an emerging research field with high potential for rapid and flexible production of therapeutic proteins. Among CFPS systems, mammalian based systems have been receiving an increasing attention due to their ability to produce post-translationally modified therapeutics. Although CHO cells are the market-dominating cell line in cell-based production systems, only little is yet known about their performance in CFPS systems.

In this study, we have achieved the successful design and implementation of an 'in-house' mammalian CFPS factory for more robust and easy to prepare CFPS using CHO cell extracts. A standardized workflow for the preparation of highly-active CHO lysates has been established and different IRES-mediated expression constructs (circular and linear) have been generated. Protein expression is optimized via EMCV IRES-mediated translation initiation and the addition of a poly-A tail for further mRNA stabilization. Using the transcriptional and translational machinery present in the CHO lysates, our cell-free factory achieves protein yields of up to 40 Âµg/ml in single 25 Âµl coupled batch reactions with continued protein production for up to 8 hours.

To address the yield limitations of CFPS, we have explored the supplementation of accessory proteins (APs) to our 'in-house' system to evaluate the boosting effect on lysate activity for improved production. Different conditions and strategies for implementing the accessory proteins (purified protein addition, transient expression) have been compared for their effect on yield, reproducibility, and cost efficiency. A strong synergetic effect on protein synthesis of both APs seems to be present with up to a 50-fold increase in expression levels of tGFP observed. Subsequently, we explored the CFPS of antibodies including Herceptin (Trastuzumab). In order to demonstrate the importance of this novel platform for commercial applications, we aim to further extend our product portfolio to "difficult-to-express" proteins highlighting the advantages of CFPS over cell-based production.

Large Scale Active-Learning-Guided Exploration to Maximize Cell-Free Production

Cell Free Systems Conference

Authors:

Borkowski, O., Pasteur Institute / INRIA

Faulon, J. L., Genoscope

Koch, M., INRA

Pandi, A., INRA

Soudier, P., Genoscope

Zettor, A., Pasteur Institute

Cardoso Batista, A., INRA

Lysate-based cell-free systems have become a major platform to study gene expression but batch-to-batch variation makes protein production difficult to predict. Here we describe an active learning approach to explore a combinatorial space of ~4,000,000 cell-free compositions, maximizing protein production and identifying critical parameters involved in cell-free productivity. We also provide a one-step-method to achieve high quality predictions for protein production using minimal experimental effort regardless of the lysate quality.

Prototyping Gene Expression in a Chloroplast Cell-Free System for Plant Engineering

Authors:

Majer, E., Massachusetts Institute of Technology

Clark, L. G., Northwestern University

Karcher, D., Max Planck Institute of Molecular Plant Physiology

Ryu, M. H., Massachusetts Institute of Technology

Bock, R., Max Planck Institute of Molecular Plant Physiology

Jewett, M. C., Northwestern University

Voigt, C. A., Massachusetts Institute of Technology

Plants are difficult to engineer due to their complex eukaryotic genomes, slow generation time and long transformation. As a result, the same genetic parts are frequently reused and very few parts are available. In addition, mistakes in design significantly impede progress because they take a long time to discover. The risk increases when multi-genic systems, which require careful tuning of gene expression, are transferred into plants. Here we apply a chloroplast cell-free system, purified from tobacco leaves, to prototype single and multi-genic expression for plastid engineering.

Plastids are plant subcellular compartments of prokaryotic origin with ribosomes that recognize Shine-Dalgarno sequences. We used RBS Calculator to design several RBSs to express GFP under the T7 promoter. Protein expression correlated with the predicted RBS strengths, suggesting that the RBS Calculator can be used to fine-tune expression not just in bacteria, but also in plastids. In addition, codon-optimized and non-codon optimized versions of two enzymes from the capsaicin biosynthetic pathway were assayed in the chloroplast extract. Errors in codon-optimization led to low protein expression. Being able to prototype in the chloroplast cell-free system saved valuable time and effort before plastid transformation.

In this work we use chloroplast cell-free extract to assay synthetic nitrogenase clusters at mRNA level, demonstrating the utility of the system for prototyping multi-genic constructs. The synthetic clusters were transformed into tobacco plastids to correlate cell-free and *in vivo* expression. RNA-seq data obtained from these plants shows that the chloroplast cell-free reflects *in vivo* gene expression.

Our chloroplast cell-free system enables precise characterization of genetic parts and the testing of 10^3-10^4 construct in a week. Constructs producing the desired expression pattern *in vitro* can be selected for transformation which will reduce the amount of work related to plant regeneration and increase the success rate of engineering complex functions in plants.

A Cell-Free Biosynthesis Platform for Modular Construction of Protein Glycosylation Pathways

Cell Free Systems Conference

Authors: Kightlinger, W., Northwestern University ,Lin, L., Northwestern University, DeLisa, M. P., Cornell University, Mrksich, M., Northwestern University, Jewett, M. C., Northwestern University, Duncker, K. E., Northwestern University, Ramesh, A., Northwestern University, Thames, A. H., Feinberg School of Medicine, Northwestern University, Natarajan, A., Cornell University, Stark, J. C., Northwestern University, Yang, A., Northwestern University

Cell-free systems are emerging as a robust technology to characterize, design, and optimize proteins and biosynthetic pathways with unprecedented control and throughput. However, the production and testing proteins with post-translational-modifications (PTMs) *in vitro* remains difficult. Protein glycosylation (the attachment of oligosaccharides to amino acid sidechains by glycosyltransferase enzymes) is among the most abundant PTMs, is present in 70% of protein therapeutics, and is known to endow protein therapeutics with beneficial properties[1]. Despite their importance, the membrane-bound and complex nature of most glycosylation systems has impeded the development of biosynthetic pathways to produce glycoproteins outside of their native hosts, particularly *in vitro*.

Here we take a modular, unit-operations approach to develop a versatile cell-free platform for glycosylation pathway assembly by rapid *in vitro* mixing and expression (GlycoPRIME). In GlycoPRIME, crude *Escherichia coli* lysates are selectively enriched with individual glycosyltransferases by cell-free protein synthesis and then mixed-and-matched to construct multienzyme glycosylation pathways. Importantly, GlycoPRIME uses a soluble bacterial *N*-glycosyltransferase to install a glucose primer that is elaborated to therapeutically-relevant glycosylation motifs. This soluble system enables the first entirely *in vitro* glycoprotein production method that does require the synthesis of membrane-associated glycosylation components in living cells, obviating the need to make new genetic constructs and strain lines for each enzyme combination. We demonstrate GlycoPRIME by screening 37 putative protein glycosylation pathways *in vitro*. These pathways provide new biosynthetic routes to 23 unique glycan motifs, 18 of which have not yet been synthesized on proteins. We used selected pathways to synthesize a protein vaccine candidate with an Î±-galactose adjuvant glycan in a one-pot cell-free system and human antibody constant regions with minimal sialic acid glycan in glycoengineered *E. coli.* We anticipate that these methods and pathways will make possible new applications in cell-free synthesis and glycoengineering.

[1]Sethurman & Stadheim, *Current Opinion in Biotechnology*, 2006

Engineering Gene Regulatory Networks in Vitro

Authors:

Maerkl, S. J., Ecole Polytechnique Federale de Lausanne

Cell-free synthetic biology emerged recently as a viable *in vitro* alternative for biological network engineering. Cell-free synthetic biology implements biological systems in a coupled transcription - translation reaction and therefore is a well-defined environment that is easier to control and interrogate than complex cellular systems. Since genetic networks can be implemented *in vitro* as linear dsDNA templates, as opposed to plasmids, it also circumvents time consuming cloning and transformation steps, enabling rapid prototyping of genetic systems. We discuss several technological and methodological advances including the development of a microfluidic chemostat device, a high-throughput microfluidic device, and a method to easily produce a recombinant cell-free system. With these various tools in hand we rapidly prototyped genetic networks *in vitro* and transplant them into living hosts. More recently we engineered gene regulatory networks from the bottom-up with synthetic Zinc-finger transcriptional regulators. We comprehensively characterized our novel parts by deriving binding energy landscapes for our transcription regulators, which in turn allowed us to precisely tune repression and optimize more complex gene regulatory network topologies. We expect that this work will form part of the technological and biological foundation required for the creation of artificial cells or cell-like entities as well as provide a simpler system than cells to study complex regulatory mechanisms in what one could think of as 'systems biochemistry.'

Robust Cell Free Extract from Bacillus Subtilis to Prototype Engineered Spore Circuitry

Cell Free Systems Conference

Authors:

Reuel, N., Iowa State University

Tamiev, D., Iowa State University

B. subtilis is a common, gram-positive bacterium capable of forming spores which our lab is exploiting to create engineered spore devices such as embedded sensors, time-delayed expression vehicles, and hydrolytic logic incorporated in advanced materials. Transformation of these cells is much less efficient than work in other model organisms, such as *E. coli*. To prototype spore-based circuitry we turn to cell free reactions. Cell free protein expression has been shown previously in *B. subtilis*[1] but we found extract made from tip-sonication batches to be highly variable in performance (much more sensitive than *E. coli* extract preparation). We demonstrate improved homogeneity in extract prepared from large-batch continuous homogenization (single piston cycle) followed by lyophilization, as we have done previously for *E. coli* extract [2]. Using this robust extract, we demonstrate the expression of beta-galactosidase and super folding green fluorescent protein. We also show utility of cell-free prototyping, by testing a panel of 6 riboswitch circuits demonstrating varying levels of translation control (expression of red fluorescent protein) when exposed to the theophylline analyte.

[1] Kelwick R, Webb AJ, MacDonald JT, Freemont PS (2016) Development of a Bacillus subtilis cell-free transcription-translation system for prototyping regulatory elements. *Metabolic Engineering*, 38:370â€"381. https://doi.org/10.1016/j.ymben.2016.09.008

[2] Dopp JL, Reuel NF (2018) Process optimization for scalable E. coli extract preparation for cell-free protein synthesis. *Biochemical Engineering Journal*, 138:21 -28. https://doi.org/10.1016/j.bej.2018.06.021

Metabolomic Characterization of Escherichia coli-Based Cell-Free Expression Systems for Process Optimization

Cell Free Systems Conference

Authors:

Styczynski, M. P., Georgia Institute of Technology, Miguez, A., Georgia Institute of Technology, McNerney, M., Georgia Institute of Technology

An important question about cell-free expression (CFE) reactions is why transcription and translation eventually degrade and stop. Commonly-suggested possible causes include exhaustion of transcription or translation substrates, exhaustion of energy sources, cofactor imbalances, buildup of toxic products, or degradation of transcriptional or translational machinery. Something in common to most of these hypotheses is that they involve small-molecule biochemical intermediates - molecules which, if they were still in a cell, we would call metabolites. To this end, we suspected that metabolomics (the systems-scale measurement and analysis of small biochemical intermediates) could be useful in characterizing when, why, and how transcription and translation eventually stop in CFE reactions. Here, we describe our efforts to use metabolomics to relate lysate preparation and performance to metabolic activity. We show that, as expected, lysate processing affects the metabolite makeup of lysates. Nonetheless, we identify some surprising trends, including multiple metabolites that are present in greater amounts in dialyzed lysate than in undialyzed lysate. We also show that, as one might expect, lysate metabolite levels change over the course of a CFE reaction. However, what was not expected was that these changes strictly due to the time of the CFE reaction dwarf any metabolite changes attributable to transcriptional and translational activity, and they occur regardless of whether a target molecule is produced. This "endogenous metabolism" of the CFE reactions is an important, poorly-characterized phenomenon that may contribute substantially to the temporal degradation of CFE. We then used our insights to devise new small molecule supplements that improve CFE protein production. These approaches and insights could be valuable in the future as more groups move towards CFE at scale, for which clear markers of lysate activity and supplementation approaches that compensate for potential variability across batches and experimental protocols would be extremely useful.

Cell-Free Prototyping for CarbonSmartTM solutions

Authors:

Mueller, A., LanzaTech

In pursuit of solutions to meet the needs of a growing population while reducing carbon emissions released in the process, LanzaTech is commercializing a gas fermentation platform. This platform enables the conversion of local, highly abundant, waste and low-cost resources including industrial flue gases as well as syngas generated from biomass resources. These gasses are converted into useful products including fuels, chemical building blocks for materials, and food. In 2018, LanzaTech successfully deployed a commercial-scale facility and with partners, has produced over ten million gallons of ethanol to date. We are actively developing our next generation of product molecules through the engineering of our proprietary acetogen. Working closely with our partners we are utilizing cell-free systems to reduce turnaround time of the research cycle, accelerating the path of these products toward commercial readiness. While existing cell-free systems based on model organisms has proven informative, we are also developing the first acetogen-based cell-free system to leverage chemistries and interrogate machineries absent in model organisms.

Cell-Free Synthetic Biology: Serving Society with Supernatural Processes and Products

Authors:

Swartz, J. R., Stanford University

Biology is amazing! Organisms and biological processes are incredibly versatile and yet, as our understanding grows, logical in their design and function. Nonetheless, the complexity of biological systems can be daunting. Cells have evolved to conquer multiple challenges and, most importantly, to be resilient self-replicating factories. But we don't need all this to provide important services and products. Cell-free synthetic biology now lets us activate and combine individual biological subsystems to focus on a single objective. This talk will begin by describing how this ability emerged through a series of foundational advances.

As modern cell-free technology emerged, we gained the ability to directly modify, monitor, and control intracellular processes. For example, in a single uniformly mixed reactor we activated synergistic processes that normally occurred in different subcellular compartments. This suggested the potential for 'supernatural' capabilities. We next activated cell membrane dependent processes to economically supply biochemical energy. When non-natural amino acids with biologically orthogonal reactivities were efficiently introduced, we then gained the ability to make supernatural multi-functional molecules and molecular assemblies.

Now Sutro Biopharma has applied and scaled up this foundational technology to make innovative pharmaceuticals such as precisely designed antibody drug conjugates. These are now showing promise in human clinical trials. Innovative biosensor technology has also been produced, and complex adaptive control networks have been activated. More recently, feasibility for small on-demand biopharmaceutical production 'facilities' has been demonstrated. This exciting development offers a new approach to pharmaceutical production and also motivates new concepts for protein pharmaceutical QA and QC. It may also facilitate globalization of cell-free biotechnologies. These are exciting times, and the talk will conclude with glimpses of new cell-free technology platforms that could enable rapid, effective vaccine responses as well as smart nanoparticles that finally provide the 'magic bullet' for precisely targeting diseased tissues such as tumors.

Engineering a Non-Canonical Redox Cofactor System for Biocatalysis

Authors:

Li, H., University of California, Irvine

Cofactors such as NAD(P)$^+$ are essential reagents in cell-free biomanufacturing. However, they represent a substantial cost in purified enzyme-based biocatalysis. In crude lysate-based biocatalysis, although it is suggested that NAD(P)$^+$ present in the crude lysate may replace the need for cofactor supplementation to reduce cost, numerous NAD(P)$^+$-dependent enzymes also present in the crude lysate may mediate byproduct formation and reducing power dissipation. Here we report the development of a non-canonical redox cofactor system based on nicotinamide mononucleotide (NMN$^+$). NMN$^+$ is a simpler and potentially less costly analog of NAD(P)$^+$. And we showed that it deliver reducing power in an orthogonal manner to NAD(P)$^+$. The enzyme that supplies reducing power in the system is a computationally designed glucose dehydrogenase (GDH) with a 10^7-fold cofactor specificity switch towards NMN$^+$ over NAD(P)$^+$. We demonstrated that this system can support diverse redox chemistries *in vitro* with high total turnover number (~39,000), including reduction of activated C=C double bonds, C≡C triple bonds, and nitro groups, as well as supplying electrons to cytochrome P450. Importantly, we also demonstrated that the protein design principles discovered here are translatable to engineering other key enzymes in cell-free biocatalysis such as glyceraldehyde-3-phosphate dehydrogenase (GAPDH) in glycolysis, phosphite dehydrogenase, and alcohol dehydrogenase. Furthermore, we established a high-throughput selection platform to enable the engineering of NMN$^+$-dependent enzymes in directed evolution. Overall, these results suggest that the non-canonical cofactors may have potential as viable alternatives in cell-free biomanufacturing.

On-Demand Endotoxin-Free Production of Protein Cancer Therapeutics from Thermostable Freeze-Dried Cell-Free Protein Synthesis Reagents

Cell Free Systems Conference

Authors:

Bundy, B. C., Brigham Young University, Wilding, K. M., Brigham Young University, Earl, C. C., Brigham Young University

Currently, shortages in certain FDA-approved protein therapeutics have put patients at risk of not receiving the drug they need. Cell-free systems has emerged as promising platform for on-demand and even on-site production of therapeutic proteins that could potential answer the challenge of protein therapeutic shortages and accessibility. The *E. coli*-based cell-free system is particularly attractive due to its low cost and robust nature. However, this system inherently contains endotoxins and here we report the use of an â€œendotoxin-freeâ€� E. coli strain to decrease the risk of endotoxin exposure from an on-demand system. The recent advent of lyophilized cell-free reagents also further simplifies on-demand applications, however its long-term stability at elevated temperatures is limited. Here we report the use of lyoprotectants to enhance the stability of lyophilized reagents and discuss the negative effects of lyoprotectants that have previously been reported for use with cell-free systems. Finally, we demonstrate the production of an FDA-approved protein therapeutic from a truly single-pot lyophilized, endotoxin-free, cell-free protein synthesis system after 45 days of storage at elevated temperatures.

Reference: Wilding KM, Zhao EL, Earl CC, Bundy BC. 2019. Thermostable Lyoprotectant-Enhanced Cell-free Protein Synthesis for On-demand Endotoxin-free Therapeutic Production. New Biotechnology. 53:73-80.

Fast and Furious: Cell-Free Biomanufacturing Approaches

Cell Free Systems Conference

Authors:

Rao, G., University of Maryland Baltimore County

Prevailing biologics manufacturing technologies are ill-suited to rapidly respond to biothreats and/or rapid fluctuations in product demand. A typical manufacturing campaign starting from a cell bank to a finished vial can take several weeks, during which time enormous casualties may be sustained. In addition, distribution logistics during a time of potential travel restrictions and quarantines would result in additional barriers to delivering the therapeutic/vaccine in a timely fashion.

We propose that these limitations can be overcome using cell-free biomanufacturing. Case studies will be presented for a rapid response to a radiological incident, where large quantities of GCSF will need to be therapeutically administered. Currently, GCSF is stored in the Strategic National Stockpile, but faces constant expiration and replenishment requirements. In contrast, we show how a cell-free approach can be used to overcome these limitations.

Other biologics/vaccine candidates, scale-up possibilities and regulatory barriers will be discussed.

Enabling Design and Manufacture of Best-in-Class Protein Therapeutics By Cell-Free Production; Concept to Clinical Trials

---

Authors:

Hallam, T., SUTRO BIOPHARMA

Use of Cell-Free protein production has several game-changing advantages over conventional cell-based expression for human therapies. Sutro has developed a microbial-based platform to enable the design and production of best-in-class complex proteins-based therapeutics. We'll provide incite into some of the fundamentally critical attributes that have guided the development of the platform that support three clinical development stage oncology therapeutics currently being evaluated in patients.

A Cell-Free Platform Based on Nisin Biosynthesis for Discovering Novel Lanthipeptides and Guiding Their Overproduction In Vivo

---

Authors:

Liu, R., Key Laboratory of Combinatorial Biosynthesis and Drug Discovery (Wuhan University), Ministry of Education, and Wuhan University School of Pharmaceutical Sciences, Zhang, Y., Wuhan University, Liu, T., Key Laboratory of Combinatorial Biosynthesis and Drug Discovery (Wuhan University), Ministry of Education, and Wuhan University School of Pharmaceutical Sciences, Deng, Z., Key Laboratory of Combinatorial Biosynthesis and Drug Discovery (Wuhan University), Ministry of Education, and Wuhan University School of Pharmaceutical Sciences

Lanthipeptides are a major group of ribosomally synthesized and post-translationally modified peptides (RiPPs) produced by microorganisms, have extensive therapeutic and industrial applications, therefore, discovery of novel lanthipeptides and increasing production are the research hotspots in the synthetic biology and metabolic engineering fields. However, since many are bactericidal, traditional *in vivo* platforms are limited in their capacity to discover and mass produce novel lanthipeptides as bacterial organisms are often critical components in these systems.

In this work, we developed a cell-free protein synthesis (CFPS) platform and demonstrated its use for discovering novel lanthipeptides and guiding lanthipeptides overproduction *in vivo* using nisin as an example. For discovery of novel lanthipeptides, we mined all nisin analogs (~210) from NCBI database, then deduplication and excluded known lanthipeptides and found 18 potential lanthipeptides, they were mined in our CFPS platform in a single day and identified 4 novel antibacterial lanthipeptides (one of them show better antibacterial activity than nisin). Next, we used antimicrobial activity screening in CFPS platform for discovering nisin mutants against the growth of gram-negative bacterial, a nisin mutant library with anti-bacterial activity was screened and achieved one mutant from 3,000 mutants with better anti-*E. coli* activity than nisin.

For overproduction of lanthipeptides, we identified NisB (the dehydrase in nisin biosynthesis) is the key system controller in nisin biosynthesis in the CFPS platform. Then the industrial nisin Z producing strain was precisely improved by target metabolic engineering *in vivo* (precursor peptide gene of nisin and *nisB* overexpression), the titer of engineered strain is 8,828 IU/ml (60% nisin Z increase than the current industrial nisin strain. This strategy (*nisB* overexpression) was expanded to enhance nisin analogs production in *E. coli,* an 89.2% increase in nisin analogs was achieved. The newly developed CFPS platform has the potential to substantially expand RiPPs research and production.

AUTHOR INDEX

AIChE
120 Wall Street, FL 23
New York, NY 10005-4020

ISBN 978-1-7138-0571-7

CURRAN ASSOCIATES INC.
proceedings.com